Peace, War and Love

by John Smale

Published in November 2008 by emp3books,
Kiln Workshops, Pilcot Road, Crookham Village,
Fleet, Hampshire, GU51 5RY, England

©John Smale 2008

The author asserts then moral right to be identified as the author of this work

ISBN-10: 0-9550736-8-5
ISBN-13: 978-0-9550736-8-7

All rights reserved. No part of this publication may be reproduced, stored in a retrieval system, or transmitted, in any form or by any means, electronic, mechanical, photocopying, recording or otherwise without the prior written consent of the author.

www.emp3books.com

This book is dedicated to Viv

NOTES AND ACKNOWLEDGEMENTS

The photographs of SS Windsor Castle are included with the permission of MaritimeQuest who own their copyright. (www.maritimequest.com)

Thanks to Commercial Campaigns for the cover design. (www.commercialcampaigns.co.uk)

A very big thank you to Jack and Sophie for allowing me to record and publish this real account of the power of love over their many years.

Contents

Introduction	1
Jack's Peace	
Growing up on a farm in Dorset 1921 to 1940	3
Photos	26
Sophie's Peace	
Growing up in Kent and Dorset 1922 to 1940	31
Photos	44
Chamberlain declares War	47
Jack's War	51
Photos	72
Sophie's War	79
Photos	98
Home Thoughts from Abroad	103
Telegrams and postcards	
Churchill Announces Peace	113
Jack and Soph Post War	115
Appendix I	123
Postscript	135

Introduction

This is a true story set in the days of black and white photographs that hid the colours of a reality that was blissful...until the War broke out. Young men and women answered the call to defend their country and to risk their lives and sacrifice their youth.

It changed people, it changed lives. Although devoid of John Wayne style Hollywood heroics, the attitudes displayed were a real sense of heroism. Young people never knew from day to day what the future held. Hope was the mainstay of dreams but the smell of death was forever present.

What was needed was a set of friends and a true love to keep smiling through knowing that they would meet again, some sunny day. And so it was, even though fate seemed to conspire against the two people in this story, it was actually on their side.

Jack had an idyllic childhood as the son of a carter on a farm in Dorset. Sophie's life started well but she found herself having to do more and more to help to look after an ever increasing number of younger sisters. As the pressure built, she left home at seventeen to live with her grandmother, close to the bombing.

Then she met Jack again. She had met him before she left Dorset, but a chance meeting brought them together again.

After courting in the Blitz they were married. Shortly afterwards they were parted as Jack was sent overseas with the Army.

During this time they were both nearly killed. They came back together after the War ended and remain that way today.

Jack's Peace

1921 was a year when many things happened that would have an effect on the World for years to come.

To name some, Mao Tse-Tung, a library assistant and primary school teacher, was at a meeting where the Chinese communist party was formed. Crown Prince Hirohito was named as the Regent of Japan. Britain signed a declaration of independence with Ireland.

In addition to all these things, Jack was born in a small bedroom of a farmhouse in the tiny village of Lake that nestled in the verdant countryside of North Dorset. He was the youngest of seven siblings. His mother was a petite lady with a big family. She was forty years old.

The sweet, leathery smell of Condor tobacco trickled from the smouldering pipe of his father. Mixing with the smoke from the log fires it seemed to set a scene for Lord of the Rings. Yet there was no magic from sorcery, only the overlaying spell of love and peace.

Jack's father was a tall man with bright blue eyes that sat above shadows that reflected his many years of long and hard work as a carter. His big black bushy moustache hid the smile of satisfaction that he got from working with his horses and the heavy agricultural carts and wagons.

He worked for a man called Perham, who rented the farm from the Digby Estate. The Digby family was descended from Lord Digby who bought Sherborne Castle in 1617.

The time of Jack's birth was a busy season on the farm, as all seasons were. The cows would soon be moved to their winter quarters as the grass became less abundant in the fields and as autumn rain would make the ground muddy. Damage from their hooves sinking into the ground would damage the grass roots and cause problems for the following years.

Fields for winter sowing had to be ploughed. Jack's father would spend his day from dawn to dusk steering the blades of the plough up and down furrows whilst controlling and encouraging his horses. This had to done so that the frosts would break up the heavy clods of earth making it easier to work with when sowing time arrived.

So, Jack's arrival into this world would have been an event, but there would have been little time for celebration.

Wheat, oats and barley would be sown using a seed fiddle, a device that would fling seeds from a bag at a given rate by rotating a handle. It had to be timed to the pace of the user. Walking too fast would not give enough seed and walking too slowly would spread far too much. This was an operation that had to be timed with precision. The turning of the handle and the pace at which the spreader walked had to be in perfect balance.

At this early age, Jack's playmates included his siblings and the animals on the farm; the sheep, cows, horses, pigs, chickens, cats and dogs. He was never lonely!

His world was contained in the local neighbourhood. There was no television or radio. The only news would be of local happenings that were discussed in the shops or in the pub, and that would be of little interest to him.

To say that life was simple would be true, yet it would give the wrong impression. There was no need for a car, not that they were affordable, because horses were waiting to be saddled and ridden when they were not pulling carts or ploughs.

Food was at hand from the vegetables in the garden and the livestock gave meat when the need arose. In his world, lack of distraction from the television or radio never meant boredom; instead it gave the chance for conversation, play and hobbies.

There was respect for other humans rather than a sense of disdain if neighbours possessed more than you. Life was about quality rather than quantity, it was that dream of escape that people in the twenty first century hope to be rich enough to be able to buy!

In this oasis of time between the First World War and the Second, Jack grew up in this wonderland of green fields, whilst Germany was gripped by poverty in its attempt to pay reparations following its first conflict.

On the farm, life was hard, very hard, but blissful. The days started early, before sunrise. Cows had to be milked twice a day. They had to be herded into the milking pens, have their teats washed and their milk squirted into buckets and then put into churns. Everything had to be spotlessly clean so washing the containers also took a lot of time and effort.

After the cows were milked and released back into the fields the men and women would sit down to a fried breakfast. Bacon, eggs and field or horse mushrooms picked from the fields, if in season. (Horse mushrooms are a different species to field mushrooms and they only seem to grow in fields that have had horses grazing on them.) There would be doorsteps of bread and butter. The food and the fatty content would be digested and turned into fuel for the manual work during the rest of the day rather than congeal, as it would today, in an office worker's arteries.

Agriculture was always a low paid occupation, but Jack's family never had the disposition to migrate into the towns to work in factories. They had sufficient food, warmth and comfort to keep them happy. But that was never in this child's sphere of concern.

Instead Jack would run through fields as they changed with the seasons. From ploughed furrows to the birth of young plants in the spring, onto mature crops and then harvest time. As he grew, the seasons rotated year after year as they always had, and he became more involved in the daily routine of the farm.

In late autumn the livestock was moved into shelters for winter and their diets changed for the cold, frosty winter days. Now indoors, their diet was supplemented with silage, grass preserved with molasses that would either smell of sweet treacle or rotting compost, depending on the sniffer's nose. Root crops were harvested and stored. Sugar beet was sent away for refining. Mangle-wurzels, big turnip like vegetables, were chopped and fed to the cows and sheep. They were also chopped and mixed with treacle for the horses. Swedes were also given to the sheep but the green leaves that grew on them were stripped off and eaten like cabbage by the family.

Everything that was used as winter fodder had to be grown and then gathered by hand. Back breaking work that involved bending and lifting.

Naturally animals kept in shelters produce waste products. In the spring and summer, they automatically spread manure back onto the land but in the darker and colder days inside, the waste had to be cleaned out of the sheds. It was washed into foul smelling slurry pits ready to be sprayed onto the fields the next year. This brushing and water throwing was a job everybody wanted to avoid, but somebody had to do it!

The animals were given timetables to follow through the year. Rams were taken to the ewes to create lambs, and then the lambs were fattened for the market. Chickens gave eggs and those that were not eaten gave more chickens. Horses gave birth to foals that were raised, sold or broken in to become loyal workers. Everything followed the natural calendar, the cycle of life. Nobody needed more.

When he was older, Jack went to the small school in the nearby village of Thornford. He had to walk the few miles there and back each day to learn to read, write and how to work with numbers. Once every week he walked to Bradford Abbas, a village five miles away for carpentry lessons. Being able to work with his hands and with tools was an essential part of life for a farmer's son.

In addition to his school lessons, two spinsters who lived in Thornford would teach him woodcarving.

Another benefit was that they had tennis courts where Jack could play. This was a joy for an active young man who would never have had access to such a middle class sport otherwise.

His three older brothers, his father and his Godfather Bill, worked on the farm, or on farms nearby in the locality. The family was part of the land as the land was much of them.

Harry, one of his brothers, worked with horses as did his father. He was a big man who was fairly insular. He had a pipe perpetually stuck in his mouth that matched, in colour, the trilby hat on his head. He always seemed to be searching for something he had lost in his world. It is said that later in his life, he had been stood up at the altar. Perhaps his forlorn waiting for his love had started before he had even met her.

The other two brothers were farmers as well. Vic, an older version of Jack, and a younger version of his father, with blue eyes and swept back black hair, had ambition and wanted to have his own farm. Fred was the eldest brother and he also worked with cows and sheep. Although he was the smallest man, he had a heart of gold and always a twinkle in his eyes. He was smart and would one day have his own farm as well.

Jack's three sisters helped on the farm until they were old enough to get other jobs. Dora, the eldest of all of them was a cook at the local vicarage; Betty would go on to work in Woolworths as did Dorothy.

In this part of Dorset the land flowed up towards gentle hills that added a sense of distance to the view. On those fields, grass grew that was rich and strong in colour and was able to sustain a strong and robust herd of cattle. They, in turn provided milk that was as thick and creamy as the accents that surrounded the young lad. It was the sort of accent that is rarely heard today. C's were pronounced as Z's. H's were missed. F's were V's, such as, "Arry be on dracter. E's on 'is way back t'varm. E neezs t'press zillerater morr." Or something like that!

The accents were part of the way of life. As distinctive as a signature, they located a person to a village. But these people were never yokels as portrayed in The Archers or any programme that includes those horrible fake West Country accents. No. They were the very salt of the Earth on which they lived and grew their living.

Jack's Godfather, Bill, was a shepherd. With a big moustache and a cloth cap perpetually on his head, he was a very practical man, to put it mildly. He would lay bricks, set hedges and would thatch houses as well as tend to the needs of the sheep, act as midwife during lambing and do any other jobs that were required around the farm.

Jack's mother, Frances, was small but fierce in a gentle way, if that paradox makes sense. She was strong. She had to be. To control big strapping farmers she needed every ounce of authority she could muster. With the respect she was given, she succeeded. The use of superstition helped. "Putting boots on the table brings bad luck" or "Spilling salt brings bad luck" Nobody would temp fate by breaking the rules. She had won control by introducing a process of influential psychology.

Bread came from the local baker who would deliver daily to his customers. Jack's mother would always cut bread in her own unique way. Standing the loaf on its end, rather than laying if flat on a board, she would spread butter on the open top, cut through horizontally with a sharp knife, and pass the slice using the knife. Nobody else would be allowed to touch the loaf. This was probably to restrict dirt-ingrained hands from contaminating the bread.

Her cooking was done on a 'range', a cooking system that used metal plates over the fire of a stove. It was powered by burning logs that were split using wedges and large wooden mallets. They had circular heads with rounded ends that were one and a half feet in diameter and with handles three feet long. These were called Beedle mallets, pronounced as 'biddle'.

Some words get lost, today. Jack's mother referred to ants as emmets and if she just wanted a small portion of something she would ask for, 'just a titsun spoonful'.

This was indeed basic, yet excellent living. She made her own butter and cheeses including Dorset Blue Vinney and a soft cottage cheese. To quote the

Dorsetshire poet, William Barnes in his poem 'Praise of Dorset':

> "Woont yer 'ave brown bread a-put yer,
> An' some vinny cheese a-cut yer?"

Once a week, the fish-man would bring fresh fish that had been caught on the Dorset coast and had never had to make the time consuming, and freshness losing, return journey via a centralised fish market as it would now! Great for people living in London, but totally pointless for those living twenty five miles away.

Pigs were only killed on rare occasions. Jack's mother would make black puddings from the blood; brawn from the head and brains, and chitterling from the intestines and everybody loved the trotters. All those things were made that fell out of, and back into, favour with gourmets. The joints were cut and hung to mature or dry.

Added to the diet there would be the occasional pigeon pie, rook pie or rabbit pie. After all, if those creatures ate the crops then they made themselves edible targets for Mr Perham's gun!

And there was hedge bounty. Crab apples, hazelnuts and chestnuts. Elderflowers, elderberries, sloes, brambles and more. The natural hedges that were managed by hand gave shelter to wildlife before they were ripped out or severely trimmed by machines.

Friday night was bath night and one after the other the sisters and brothers would be immersed into water from the well that was heated in a big copper boiler that was also used to boil clothes clean.

Jack's mother and father would wash upstairs using a jug of water and a big china basin. With a po under the bed, this was the equivalent of having an en-suite bathroom, today. The everyday toilet was a cold, dark shed outside.

His father, Louis, cleaned the boots for the children and for his wife. In return she would always make sure the washing up was done before going to bed, 'in case the doctor has to be called out!'

Much to Jack's mother's annoyance, his father used him as weight on the backs of big Suffolk Punch horses when breaking them in. Jack learnt to ride horses that way, and they became the first love of his life. He would be placed on their backs and trotted round in a circle, Jack's father holding a rope. To a child this was seventh heaven like a wooden fairground horse but less predictable.

Those horses were the heavy machinery of the farm. They were used for pulling ploughs, mowers and wagons. They would be hitched to ploughs with leather straps and metal chains and the laborious business of walking up and down fields would start. Jack's father would follow at the rear encouraging the horses and steering the plough or whatever else was attached. His view would be that of tails whisking away the flies. Birds would flock around in search of worms and bugs that had been uncovered by the plough. Days for farmers and horses were long and hard.

Following the hard work of gathering the crops through the autumn and early winter, and before seeding and planting out started again, Christmas was a marked break. It was a time for a big family get-together. As a farmer it is difficult to take time off. The animals still need looking after, being fed and milked, but presents would arrive magically during the night of Christmas Eve. Stockings were hung and then filled with oranges and nuts. In addition there would be books or 'annuals' and a few toys.

Christmas dinner was chicken. Real chicken that was raised and fed for the occasion. They had plump white breasts and brown flesh on the legs.

They were never the limp and bland supermarket things that are eaten today with all white meat that need Tandoori powder to give them any flavour. Those chickens had lived a free life pecking around for scraps of food in between meals of corn, rather than being jailed from birth in small cages. At night they would be put into runs to keep them safe from roaming foxes.

Along with the chicken there would be lots of vegetables. After eating Christmas dinner there would be friendly squabbling for the wishbone before a huge Christmas pudding was served with lashings of cream and custard. Organic food was never a term of description; it was a natural way of life. All this was washed down with water from a well that was clean, bright and fresh. The men would drink beer or gin to celebrate this landmark in the year when the nights started to become lighter and the weather warmer.

One day Jack's father was leading a team of horses pulling heavy logs on a wagon. A milk cart appeared from around a corner on the road and the clanking of the milk churns, metal cans three feet tall and eighteen inches in diameter with lids like inverted bowls, frightened the horses and they panicked.

Rearing up and then trying to bolt, they pulled the wagon over Jack's father's legs. The big iron clad wheels bearing the weight of the wood broke one leg in one place and the other in two places and left his limbs limp and crushed. The milkman rushed to the town to raise the alarm and to have the injured man taken to hospital. Jack remembers visiting his father in hospital. He was eight years old.

There was no medical insurance, no Welfare State. If a man was unable to work, he received no wage. The rest of the family had to work harder to keep the farm running, probably for no extra money.

Yet neither ever lost their love for horses. His father with his bright blue eyes still shining, although walking with a heavy limp, continued to farm and work with them until he retired. He loved his horses so much that he would sleep with them in the stables if they were a little poorly or if he felt he needed some space in his life.

Jack had his own horse called Polly. She was too wild to do farm work. Yet, that was the essence of the place. Rather than get rid of a horse that was unfit for work, it was considered to be something to cherish, love and enjoy.

The seasons still dictated the work. Although Bill was a shepherd, the sheep were sheared by the farm manager, Mr Perham and his son. The wool would be put into special bags and then taken away on wagons to be spun.

In the spring into early summer, the lambing season meant long nights spent with the sheep, tending to their needs and helping the lambs when they were born.

Sometimes they were born in bag-like pouches from the placenta that needed to be pulled away from the face to prevent suffocation. This was also the time of year when crops were planted.

It was also when the fields were fertilised with the slurry to help growth. The lambs were fattened and salt licks were put down for the animals to take in extra minerals.

Clover was a friend and a foe. It is able to 'fix' nitrogen from the air so that the need for chemical fertilisers was unnecessary, even if they had been available. However, if the sheep or cows grazed on it to excess then it would produce wind in their abdomens that would make them swell up like balloons. They became 'blown'. The only way to save them from dying was to use a metal spike that was held inside a metal sheath. This would be stabbed into the animal's stomach with much force. Then the spike was withdrawn and the gas escaped. Sometimes they would live, sometimes they would die. The stench from the methane gas was awful. Needless to say, before this was done, pipes were extinguished!

In the late summer, hay had to be made and stored.

Before tractors came along, the horses would pull mowers to cut the grass that would be dried and used to feed the animals over winter. This was then raked over and over so that it would dry in the sunshine. The rakes had big wires wheels that span and whisked the grass into the air before it settled into rows. After it had dried it had to be collected and stacked into ricks, those big house shaped blocks that had to be protected from the rain with canvas. The ricks were placed to be near the livestock living in winter sheds so they were usually at a distance from the fields where the grass grew.

During the hay making seasons, Jack had the job of leading the other horses that were pulling wagons loaded with hay up to the ricks, and then bringing the empty wagons back to the fields. He would be told off if he trotted the horses. His father would say, 'They've had a hard days work. Walk them!'

One day during haymaking Jack was standing and holding Harry's horse, Black Beauty, by the reins. The horse was hitched to the hay wagon. He was waiting for the hay to be loaded by the men scooping up the huge piles of dried grass using pitchforks. He looked down to see the zigzag markings of a huge adder slowly slithering around his foot. It then moved on to do the same around the horse's hoof. Both Jack and the horse stood stock still until the snake, tongue flicking, nonchalantly made its way into a ditch looking for smaller prey.

Gaiters were those stiff rolls of leather, buckled at the rear, which protected the lower leg down to the boot laces against thorns, snow and other dangers. On the farm they were not for decoration or a substitute for Wellingtons, but they were also against the chance of stepping on a snake, far more plentiful in those fine days where creatures had not been poisoned to death by the application of pesticides, herbicides and nitrates.

The absence of chemicals allowed life to flourish in a different way. Butterflies fluttered to parade their colours around the fields. Bees would find enough flowers to produce a surfeit of honey. Sweet smelling wild flowers abounded such as primroses, cowslips and blue bells. Poppies brightened the fields and cornflowers grew among the crops.

Harvesting was done by using a machine drawn by horses, called a binder. It would cut the wheat, barley or oats and the crop was tied into 'stooks', little wigwam shapes, to dry. The ears would be collected and the straw would be used for thatching. Nothing was wasted.

Through his childhood and beyond, one great pleasure was Pack Monday Fair. It is held in Sherborne, even to this day, on the first Monday after the 10th October, Michaelmas Day, (which was September 29th until after the calendar reform of 1752!)

Traditionally, in earlier times, this was when agricultural workers would look for a new work for the following twelve months. Servants and farm labourers were hired from one October to the next one and they would dress up to show the occupations they currently had or wanted. Maids (a word also used by

Jack's mother as a term for young girls) would carry mops, shepherds would show their sheep and so on. Some believe that Pack Monday is a mispronunciation of the word 'pact' as a result.

Others believe it was to commemorate the end of the building of the Abbey in 1490, or its later repair, and is about the builders packing up their tools. The Fair starts with a procession through the town when people make as much noise as possible to frighten away the devils by blowing whistles, cows' horns and banging away on saucepans. This crowd is known as Teddy Roe's Band. Teddy Roe was, apparently the foreman of the builders who repaired the Abbey after fire damage.

The showing-off of skills with animals may be the reason why Jack and Bill would decorate the sheep with yellow powder before taking them to the Fair for sale. They were shown off first and then auctioned. The breed of sheep was the Dorset Down with big pretty black faces. Those stocky sheep must have been a bizarre sight with their yellow wool.

Their horses were decorated with ribbons in their manes and on their tails. The streets were packed with market traders selling china, brandy snaps and fruit.

Gipsies would parade their horses through the streets, trotting them up and down. They would be sold on a handshake.

Jack, as young as he was, and the other participants and spectators would carry on to get merry as the evening came. There were boxing booths and a Wall of Death all against the noise of steam engines powering the fairground rides.

Amy was an Old English Sheepdog, probably named after Amy Johnston, the flying pioneer. Bill wanted a new sheepdog and he was talking to other men in the pub. One said he knew of a dog that the owner wanted to get rid of. So Jack and Bill walked to the house where the dog was kept. Like Jack's father the dog had been the victim of a milk-float. She had been run over by one and now had five silver ribs. She followed them home and then followed Jack everywhere. Later, when Jack came home on Army leave Amy knew of his arrival long before anybody else and would run down the road to greet him.

The entertainment would revolve around village dances where his father would play the concertina after being reminded of tunes by Jack's mother.

Yet against all this amusement was the undertone of a War coming.

Jack and his family heard the news of the build up to War on a radio that was powered by an accumulator, a type of battery that was taken away to be charged with acid. There was no electricity in the house. Lighting came from lamps and candles. Pest control was the responsibility of the cats.

In the years since he had been born, Wall Street had crashed as had the British economy. Hitler had come to power was building military resources. Fascism had caused a Civil War in Spain and Mussolini had become the Fascist dictator of Italy. Austria was also a Fascist country.

Further abroad, Communism had grown in Russia, and Japan was at war with China. The War to end Wars had, rather, seeded a period of unrest that had led to disorder, confusion, bigotry and hatred, worldwide.

Jack left school but did not want to make his career on the farm although he still helped with the work. He did haymaking, and helped with the lambing on Saturday nights when Bill was in the pub. He also assisted in looking after the other animals.

After leaving school, he got a job as a gardener which he hated. The man he worked for was unpleasant, critical and generally a nasty person so he left there and went to another place, also as a gardener.

Shortly afterwards he left that job and went to work in the stockroom of Woolworths in Sherborne. Much to his delight, he was the only male there with all the girls. He was now just 17. His boast was that one weekend went out with five different girls. Life was more innocent then, however.

The War started on 3 September 1939; Jack was just 18.

Jack worked in Woolworths until 1940. During his time there he went out with most, if not all, of the girls including Marie, an evacuee from Lambeth, London. Jack has often said that he took her home with her sitting on the crossbar of his bicycle. After telling this story, he would add, with a wink in his eye, "but it was a girl's bike!" It was a joke.

It was in Woolworths where he first met Sophie, the woman he would later marry, but they did not start courting then. That would come later.

With the War now in full swing, he left Woolworths and went to work building aircraft hangars in Watchfield near Swindon. This was a wartime airfield built specifically for the purpose of training the RAF. After he had finished there he moved down to Yeovilton to do the same thing for the Fleet Air Arm until his contract ended.

Put simply, there was no other work so he and his friend Stan Foot, joined up. He wanted to join the Air Force as a rear gunner on a Lancaster but he was persuaded to join the Army instead, which is just as well because rear gunners had an operational life of three to seven raids before being killed or injured.

The window of the room in which Jack was born

Jack as a lad with his father's concertina

Jack, the heart-throb!

Dorset Down sheep and Bill

Sophie's Peace

1922 was a year when many things happened that would have an effect on the World for years to come.

Tutankhamen's treasures were exposed for the first time in 3000 years. Germany was in dire financial straits. Mussolini became the Fascist Dictator of Italy. Gandhi was jailed on charges of sedition.

In addition to these events, Sophia was born in a hospital in Sevenoaks, a large town in Kent. She was the eldest daughter and so, in keeping with tradition, was named after her mother. She was also given the middle names of Julia Rose. Her mother and father would, from then on, add a flower's name to all their daughters.

Her first recall is that of standing in someone's living room. There was a pram which contained her younger sister, Minnie. She also had a small toy pram alongside. It was Christmas time. She had a new pair of red Wellingtons as her present which she would constantly wear while she held her toy cat, Felix.

Her father, Robert, was away from home a lot. He was a railway train driver, perhaps the equivalent of being an airline pilot today. At the age of 15 he had lied about his age to join the Army during the First World War. He was sent to France to fight in the trenches where he lost the thumb on his right had, "A nasty

German shot it off!" is all he would say. His tongue was affected so badly by mustard gas that he could drop pins through the holes. He was invalided out as a result of losing his thumb, the gas damage and trench foot. He was just 17.

During her childhood in Sevenoaks, Sophie would walk up a hill from the family house to see her gran and granddad who she loved very much. She was still a very young girl when her father was transferred to Brighton, where they now lived in a big house that gave her a sense of grandeur and comfort.

After a few years she went to the school that was along the road. On Sundays, although their parents had little faith, she and her sister went to something called Daniel's Band, a Sunday school.

In Brighton, it was a great delight to be able to walk to the seafront. There she would watch the waves rolling onto the beach and listen to the pebbles rolling in the surf. If she had been to Sunday school in the morning, then in the afternoons she and her sister would sing all sorts of hymns. Her favourite was "All Things Bright and Beautiful."

At that point, her life was also idyllic. There was a time when her grandfather, the man who ran the local graveyard, took down her a freshly dug grave hole. Even in the daylight she was, to her great surprise, able to look up and see stars. She made her wish.

She did not go to school until she was six but she able to read and write because her mother had taught her.

She was top of the class before she had been there for a term. Her mother's wish was that her children were well educated and what they did not learn at school they would be taught at home. Whilst in Brighton she took her 11+ and passed, but her family then moved back to stay with her grandparents back in Sevenoaks for a while. The reason was that her father had decided to change locations completely for some reason.

The family moved to Bradford Abbas, a small village in Dorset, just three miles from Thornford. The news arrived that informed her that she had won a scholarship. It was arranged to be transferred and her mother took her over to Sherborne to sit her entrance exams to go to Lord Digby's, the local Grammar School.

The house the family had was called "The Old School House." Memories of the house's previous role were evident. There were pegs in the kitchen where the pupils would have hung their coats. Dating back many, many years, it was a big old house with plenty of room. They had not been there long been there long before her mother said she had seen a ghost of a monk walking through the passage. She had said, "Good morning" and he had just smiled and disappeared. At least that is what she said.

Another time she was going up the stairs and he was sitting in a little recess at the top of the stairs. The sisters always joked about her ghost sightings.

The ghosts might have perhaps, been inspired by her father. He would come home with soot from his railway engine over his face, around his nose and eyes. The smells of coal smoke and steam buried themselves into his clothes. He would leave his train at Yeovil Junction where another driver would take over. He would then walk the three miles home beside the railway track that ran only a few hundred yards from the house.

Meals for the household were healthy. There would be roasts on Sundays followed by shepherd's pie on Monday and so on until the joint was used up. Sometimes there was fresh fish from the fish man who delivered on Sunday mornings before the visit to Church. Groceries were ordered from a man from International Grocers and then delivered to the house. This was the equivalent of on-line shopping today. Very little is new.

Water was pumped from a well that was just over the road. Milk was delivered in milk churns and ladled into jugs. This was used to drink and to make the morning porridge.

Eggs were delivered from the local farm and bread was bought from 'Patch the Baker' who used to bake his bread in huge ovens in the village. It seemed that everything was brought to the house. With a growing number of children, Sophie's mother was not over willing, or able, to go out to the shops.

Water was heated up in the 'copper', a big cauldron in a room beyond the kitchen that sat over a fire. This was where the washing was done. It would then be squeezed through a 'mangle', two cylinders of wood that were rolled against each other with a handle. This removed water from the clothes before they were hung out to dry in the fresh air.

Bath night was special. Water from the copper was carried into the lounge and poured into a tin bath. The girls were washed and dried with big towels. After the bath they were wrapped in their dressing gowns and carried up to bed. In the winter the cold beds were warmed with a warming pan. Sometimes a hot brick wrapped in a blanket was used or a pottery hot-water bottle.

There was fruit from the garden, Sophie's father's pride. Eating apples, cooking apples, giant blackberries, blackcurrants and gooseberries abounded. There were walnuts from an old tree that still lives in the garden today. Sometimes rabbits and bags of chestnuts would arrive at the door addressed to "Driver Bob". Her father often drove his train to Southampton where he acquired bananas, things that the other children at school had never seen or heard of. He would also have been very aware of the rumblings of another war coming later on. His world was a large one where he would meet, and talk to many people.

He grew his vegetables to perfection. Gardening for a farmer is a busman's holiday, yet to a train driver who spent his working days in a hot cab, gardening was an escape. His secret was double trenching the contents of the outside toilet into the soil! Deep enough to be far away from human contact, at least four feet, it would give manure back to the plants.

The earth would be dug down for two feet and then trenches would be dug down another two feet in depth to resemble the lower part of a staircase. After the 'manure' was put in, it would be part covered with earth to keep flies and other pests away. This continued along until the lowest trench was completed. Both trenches would be filled back to ground level with fresh earth from the next generation of trenches dug alongside. Sophie's father was a true recycling man. But it was even stranger that he loved digging trenches even though he had nearly lost his life in them during the Great War.

Sophie and her sister Minnie had their own areas of the garden to tend. The work was demanded and failure to fulfil their duties was punished by a stern lecture. One day they were given the job of planting small out cabbage plants but were distracted by their need to play and to let off energy.

Sophie's mother became aware that they had not done their work and told them to plant the cabbages out before father came home. They had to do it by candlelight to avoid being told off. Somehow, apparently, that was the best crop of cabbages that was ever grown there and only a few people knew the secret!

Sophie's Christmases were always a surprise. Its arrival would not be announced. The girls were sent to bed on Christmas Eve and told to hang their stockings from the end of the bed. There were no decorations, no tree. However, their stockings would be filled with oranges and nuts.

They would go downstairs in the morning to find that the lounge had been transformed into a magical palace. There were decorations everywhere and a tree covered with balls and candles. In addition to her stocking presents the girls would get things like knitting kits that included needles and wool. Sometimes they would get little necklaces to wear. This surprise never seemed to wear off.

Once a year the family would travel up to London and visit the Caledonian market to buy, for the girls, either a cardigan or a jumper. It was a great treat.

Every so often, every two or so years as it turns out, she and Minnie would visit their grandmother. Each time they returned there was another baby in the house. They would travel by train 'under the care of the guard' to Waterloo, then across to Charing Cross and on to Sevenoaks, again, 'under the care of the guard' to be met by their grandfather.

Sophie would help to care for her growing number of sisters and would do her homework and study. Learning was the spirit of the house. Her mother spent hours reading anything and everything to devour knowledge and information. As a result Sophie had a childhood devoted to a lot of housework. There was little freedom to do the things that the other girls were doing.

Whereas Jack was the youngest and therefore grew up with siblings leaving home and making more space and a smaller demand on financial resources, Sophie was yet to have a total of six sisters, all younger than herself. This would cause time and money pressures, and as she was the eldest, she was expected to work in the house as a cleaner, a carer and an extra provider.

She had to leave school when her mother was expecting yet another baby. She was needed to look after her sisters and to provide extra income for the family so she got a job in a hat shop, but hated it.

Fate played a hand. She was offered a job at Woolworths.

She had not been there long before they made her a supervisor. One of the girls there was called Betty. She was like a big sister to Sophie.

One day Betty asked Sophie to go to her house at Thornford for tea, but Sophie's father would not allow it because she was not allowed out if it was dark. Betty told her not to worry because her brother would take her home. That was perhaps the same as telling a chicken that a fox would tuck it into the run!

That is when she met Jack. Needless to say, they actually knew each other already because they both worked at Woolworths. However, Jack was a gentleman and took no advantage. They did not go out with each other at that point and just became friends.

Whilst they were working at Woolworths the Second World War started. Even in the remote town of Sherborne, the effects were felt. One day there was an air-raid. Perhaps the German navigators mistook the town for Yeovilton or another military base, but there was a pointless raid on the town that made a terrific mess. Jack had a cousin who was the mother of six children. In that raid she and all her children were killed.

Jack went off into the Army and Sophie, as mentioned before, was discontented at home because her father was so strict and she had become resentful of having to look after her sisters. Her father forbade her to join the Forces. His experiences in his War had been enough to stop him from his daughters risking their lives.

On many occasions before she left home she could hear the bombers going over to bomb Bristol. She could see the sky being lit up with the light from the flames. She was made very aware of the War even though she lived in a quiet village.

Her imposed role in the house became too much for her and to escape from the parental home, she asked her grandmother if she could stay with her, and she agreed.

While she was back in Sevenoaks, her mother's sister also moved in. She was just five years older than Sophie and had divorced her first husband because he had been unfaithful.

They both got jobs in Sevenoaks, a town that is to the south of London and was often the target of bombs that fell short. The town became a target for bombers flying to bomb London and for dumping bombs on the way back. It was not a safe place to be in. Perhaps being in the Forces would be safer and less frightening.

Later in the war, Sevenoaks was one of the towns that became a target for doodle-bugs. They were unmanned flying bombs that flew over from France until the engine cut out. Then the silence was terrifying as people waited for it to explode never knowing where that would be.

One night Sophie's young aunt was coming home and met a fresh faced Welshman. He was in the Army, however, and shortly after he was posted. That was the way of the world. (However, their love blossomed by letter and she married him later on.)

The two young women decided to join up. They both decided to join the Navy but there was a waiting list so they went to the RAF recruiting office and were both accepted. Those little twists of fate run through the lives of everybody.

She joined the Air Force just two months after Jack had joined the Army.

Sophie with her red 'wellies' and Felix the cat

Sophie's mother and father on Brighton beach

Sophie's father, 'Driver Bob' at Yeovil Junction (after the War)

Prime Minister Chamberlain declares War

At a little after 11.00 on 3 September 1939, Neville Chamberlain broadcast the following speech:

"I am speaking to you from the Cabinet Room at 10 Downing Street. This morning the British Ambassador in Berlin handed the German Government a final note stating that, unless we hear from them by 11 o'clock that they were prepared at once to withdraw their troops from Poland, a state of war would exist between us. I have to tell you now that no such undertaking has been received, and that consequently this country is at war with Germany.

You can imagine what a bitter blow it is to me that all my long struggle to win peace has failed. Yet I cannot believe that there is anything more or anything different that I could have done and that would have been more successful.

Up to the very last it would have been quite possible to have arranged a peaceful and honourable settlement between Germany and Poland, but Hitler would not have it. He had evidently made up his mind to attack Poland, whatever happened, and although he now says he put forward reasonable proposals which were rejected by the Poles, that is not a true statement.

The proposals were never shown to the Poles, nor to us, and though they were announced in a German broadcast on Thursday night, Hitler did not wait to hear comments on them but ordered his troops to cross the Polish frontier the next morning.

His action shows convincingly that there is no chance of expecting that this man will ever give up his practice of using force to gain his will. He can only be stopped by force.

We and France are today, in fulfilment of our obligations, going to the aid of Poland, who is so bravely resisting this wicked and unprovoked attack upon her people. We have a clear conscience - we have done all that any country could do to establish peace.

The situation in which no word given by Germany's ruler could be trusted, and no people or country could feel itself safe, has become intolerable. And now that we have resolved to finish it I know that you will play your part with calmness and courage.

At such a moment as this the assurances of support which we have received from the empire are a source of profound encouragement to us.

When I have finished speaking, certain detailed announcements will be made on behalf of the government. Give these your closest attention. The government have made plans under which it will be possible to carry on work of the nation in the days of stress and strain that may be ahead...

Now may God bless you all. May He defend the right. For it is evil things that we shall be fighting against - brute force, bad faith, injustice, oppression and persecution - and against them I am certain that right will prevail."

Jack's War

Almost exactly one year after the Second World War began for Britain and a few days after his 19th birthday; Jack volunteered to join the Army with his friend, Stan Foot. They enlisted on the 16th September 1940, three months after British troops, the remainder of the British Expeditionary Force, were evacuated from Dunkirk with appalling losses. Jack's brother-in-law, his sister Dorothy's husband, was just one of the soldiers who was badly injured. He had been found on the beach when the dead bodies were being collected. He had moved a little and was discovered to be alive. He was brought back to England to recover, but he remained a frail man for the rest of his life.

Jack went to Colchester with the Royal Warwickshire Regiment where he undertook his six weeks initial training. Although he was trained by the Calvary, who still had their horses, it was now all about metal and armoured machines.

After training he was posted to Hackney, in East London where he did a three month course as a coppersmith. This was during the height of the Blitz which had started on the 7th September and continued for months.

One night two of Jack's mates from the civilian billets where they were staying asked him to spend the evening in a pub. He went with them. Music was playing on an old piano, men and women were milling around chatting and drinking. For no reason whatsoever, one of them felt uneasy and said, "Let's get out of here." So the three of them left. Shortly after they had gone, the pub took a direct hit by a bomb.

That was the gamble of life during the Blitz. It was a horrible and frightening thing for him and other people who were there. People would hear the bombs coming down. They whistled as they fell. Nobody could tell where they were going to land. They could land where the hearer was, or up the road. They could land anywhere. The explosions would be heard and felt all around, anybody could be killed. There was a randomness about the bombing. On a battlefield soldiers knew where the bombs, shells and bullets came from. In a city, they were dropped without precision, without a specific target. They were dropped to kill, maim and frighten.

The spirit of a Nation comes to the fore in such times. It was necessary to live as though nothing would change and that fate would be a guardian angel.

From the middle of November, the mass bombing raids became unpredictable. Some nights the Germans missed London and concentrated on cities such as Coventry, sometimes industrial centres and ports.

As if to point two fingers at the Luftwaffe and Hitler, when London was not targeted, Jack used to go and sing in a pub for free beer. Using a basic microphone and accompanied by a man playing the piano, he would sing War songs that were the hits of the time. "There'll be Bluebirds over the White Cliffs of Dover" and "Don't Sit Under the Apple Tree with Anyone Else But Me." Songs that heralded victory, and songs that gave jealous warnings to suitors of loved ones.

Being 19, and with a history of liking female company, it was inevitable that he would meet a girl there. He went out with her for about three months. She would give him twenty Players cigarettes every night. "Wanted to get rid of me, I think." Jack would say, laughing. Cigarettes were considered to be 'cool' then, rather than the hideous killing tubes that they really are.

She was actually engaged to another soldier. Whenever he came home on leave she would run down to tell Jack to stay away. One day, Jack was sitting in the pub when she walked in with a big, a huge, man. Jack stared across and silently shivered. This man was big enough to hurt him really badly.

He slowly walked across the pub to where Jack was sitting. He looked at him in the eyes and said, "I'm Ivy's brother in case you wanted to know who I was."

After that, Jack got on well with him. This was a part of the 'live-for-today' spirit in the War. The girl returned to her fiancé after he was posted back to

London.

Jack was then posted to Loughton in Essex with the first ack-ack (ant-aircraft guns) workshops. After a while he was transferred to the Royal Army Ordinance Corps and posted to Pembroke Dock in South Wales. The bombing seemed to follow him. One Saturday evening, just after the bombing had started, a young couple who had just got married were staying in a hotel just down the road. It took a direct hit and they were both killed. They had been married for only a few hours.

The bombing was so intense that his unit was moved to Haverford West with the First ack-ack workshops. Once again they were billeted in civilian quarters.

One night there he went out drinking with his friends. He got so drunk that he fell between the wall and his bed. The big thump and laughter woke everybody up. Jack fell asleep where he was. The next morning the landlady wanted to know what had happened. Her grandfather, who also lived in the house, wore a nightcap because he had a big ganglion on his head. She said "Grandad was coming in to you". The frightening image of this old man, like a Dickens's ghost in a nightgown and nightcap, wobbling into the bedroom of young drunken soldiers while shaking his walking stick, started the laughter all over again.

Earlier, one of Jack's mates had been on guard. He

heard a noise, challenged it with no reply. Instead of an enemy soldier, a horse trotted by. Fear, shock and relief were ingredients of the awakening of young men as they became soldiers of War.

While he was in South Wales, he was given leave and went home to Dorset. He went to a dance in Thornford and, by chance, Sophie went to the same dance. Jack never liked dancing and he had just wanted to go to the pub to have a few beers and chat up the girls. He re-met Sophie and asked if he could meet her the next morning at Sherborne station. They finished up in the City of Bath and went to the cinema and watched Chatanooga Choo Choo starring Glen Miller and Sonja Henie. They became a 'proper couple' after that and Glen Miller's music stayed with them for ever.

They were now a partnership. Sophie would travel to South Wales to visit him now and again when she had leave, or he would travel to London to see her. They married in August 1942.

After Haverford West he was posted to Catterick to be in the workshops, working on tanks. He was now in the REME (The Corps of the Royal Electrical and Mechanical Engineers that was formed on 1st October 1942).

However, Jack hated Catterick, not because of the place or the work, but because of his Sergeant Major.

It started when Jack had travelled from South Wales for the first time. He had spent all night on a train that was avoiding air raids. He had his breakfast and was taken to his quarters. As he was now married, he was given married quarters. The Sergeant Major shouted at him and told him to get out and he put Jack on guard duty that night. That started a battle of wills between the two men.

However, shortly after his posting to Catterick, he was sent to Aldershot where he waited to be posted overseas. He knew he was going to be sent away on active service but had no idea where that would be. Whilst in Aldershot the training continued, and this time the training was more about fitness and discipline. So, when route marching from Aldershot to Hook with a full pack, Jack and his mates would go and hide in the bushes and smoke cigarettes.

In contrast to the Catterick Sergeant Major, there was Sergeant Major Parker. The men called him Dick but not on parade. One morning they went down to the sergeant's mess. They said they were late because they could not get up in the morning. He said, "No, don't worry. Have a lay-in. Come down when you're ready." Apparently, he was a bit like Sergeant Wilson in Dad's Army. A gentleman. He knew that these young men were going to be sent away soon, and he understood that many would never return. He had no need to assert his authority on them.

The troop was boarded onto trains and taken to Greenock in Scotland. They had no idea where they were going. The young men climbed the gang-planks onto a huge ship carrying their grey kitbags on their backs. Two funnels smoked in the early morning light. This was the S.S. Windsor Castle, an ex-mail ship/cruise liner that had been converted to carry troops. Slowly the ship sailed down the Clyde and met the cold sea.

This was Tuesday the 16th March in 1943. By then Jack had been married to Sophie for five and a half months. The ship the men were on was the SS Windsor Castle.

After the cold weather in Britain, the warmth on the ship was a comfort, in the beginning. The weather outside became warmer as the ship sailed southwards, the cramped conditions in the cabins full of soldiers, became slightly claustrophobic.

The men, amateur navigators, must have been plotting progress but the destination was only known to the crew and the escorts.

The sun shone brightly at first and lightened the apprehensive mood of the passengers. Days were spent wandering around and practising boat drill and the nights were spent in large cabins that had been sealed and blacked out to avoid showing lights to enemy planes or submarines. The boat drills, the conduct of the men in case of emergency, were carried out to ensure that everybody would know what to do if there was an attack.

Only the crew was aware of passing Gibraltar.

Dressed only in pants and a vest, Jack was in his bunk, asleep in the warm cabin. This had been a good experience so far. Sailing on this great converted cruise ship was better than route marching.

Then there was an explosion. The mood changed. The men jumped from their bunks in startled uncertainty about what had happened. Sirens screamed. The ship shuddered and slowed.

That Sunday, the 21st of March 1943 at 3.00 a.m., the ship had been torpedoed by an Italian plane.

As part of measures to avoid panic, the troops were locked into their cabins that were crammed with 20 or 30 men. After the lights went out, the cabins were pitch black. The windows had been sealed and blacked out to avoid showing the ship's location at night.

In the darkness, the men were given the last rites by a Roman Catholic priest who was also locked in with them. In that finality of gloom and fear Jack had the sudden thought that he had left his wife pregnant. He was right.

Eventually, they were released as their turn to evacuate came. The cabin was unlocked and the men made their way to the deck. Jack only had his vest, pants and socks on, plus a pair of trousers that he had pulled on. Boots might have weighed them down into the water if they were being worn. All other belongings had to be left behind, including photographs and letters. They had to jump from the high deck of the liner onto the low decks of the destroyer that sailed alongside. Jack, at that point, could not swim. As he said later, "It made no difference. I couldn't have swum 90 miles to Algiers, anyway!"

The drop was a long one and had to be timed so that as the ship tilted to one side and the destroyer rose in the water, the drop was less far. Jack saw one WREN (Women's Royal Naval Service) who mistimed the jump and fell between the Windsor Castle and the destroyer. As the boats came back together she was crushed. It was like watching a horror occur in slow motion.

His jump was made and he crashed down onto the deck of the naval ship alongside. Once on the destroyer, the men had to help find and bring in troops who had fallen into the sea from life rafts. Jack helped to pull one man on board using grappling nets. He was extremely cold when he was brought on board. He died a little later from hyperthermia. His body was buried at sea.

The destroyer delivered the men, (others had been rescued by other destroyers) in Algiers. This was the first that the troops knew of their destination. None knew what the state of the War was in that place.

As a result of the ship being torpedoed and the subsequent emergency evacuation, in Algiers the troops had no kit. So they had to walk around without shoes or boots. They were in just their socks waiting for new equipment to come.

Their accommodation was under canvass, so tents were erected that would sleep eight or nine men. These tents were pitched on the seashore of the Mediterranean. This would have been a superb holiday location, but in War, and with the added hazard of strafing from German and Italian planes, canvas was the worst form of protection.

One day Jack was walking along the beach and found a sheet of paper laying just above the water's edge. It was from the log-book of the Windsor Castle dated and 1st April 1934 after she had sailed from Mossel Bay in South Africa. This was a strange coincidence as the records of the ship were said to have sunk with her and this was a beach where the troops she had carried on her last voyage, had landed. It was if part of the ship was fulfilling its duty to see the safe disembarkation of its load. Come what may it was an omen of good luck. It is odd that the page was there. The Mediterranean is not very tidal, so there was always a sense of mystery about its discovery.

He had now been transported from a country that he knew well to a different continent, a different climate and a different culture.

Algiers must have been such a breath-taking experience. It was Jack's first time overseas and he had landed in a city with white buildings that were different in appearance to anything he had seen before. They had been built to cope with hot sunshine and different living cultural styles and set among palm trees. The markets were unaffected by the War. People were selling food and gifts to this new crowd of customers who would be eager to buy souvenirs.

Smells were alien to him and although not allowed to mix with the Arabs, aromas know no boundaries. He would smell garlic cooking in olive oil with the scents of spices. The predominant smell was of coffee.

He knew he was a long way from Dorset. The Aurès Mountains, part of the Atlas Mountains provided a backdrop and also a barrier to other parts of Africa. The Mediterranean made a beautiful blue barrier to Europe.

As if to add to the dichotomy between War and the first visit to a hot country, showers were rigged up. After all, these men were engineers and nothing was beyond their skill and initiative.

In North Africa Jack found that there were three different influences; namely, the work, the entertainment and the social chaos involving the local inhabitants and the Vichy French.

Jack's work was restoring the engines of the Army vehicles to good order. This was a vital part of what was going on in the fight against the German troops although the War was moving over to Tunisia where the Germans surrendered in that May. Algiers was just a few hundred miles away from the action.

The vehicles and engines were being returned to be mended after the ravages of fighting with enemy soldiers and the sand.

He worked in hot, dirty and greasy conditions. Whereas the work gets fewer words in comparison to the more dramatic events, it was long and hard. Being uneventful there is little else to relate, but the efforts of these men should never be underestimated.

One of his other duties was to go on patrols, in pairs, on the outskirts of the Arab quarter. Marching around in uniform, the matt REME badges on their caps, they were given the task of keeping roaming soldiers out of the Arab quarter. Soldiers mix with civilians suitably in the right place, but a drunken and maudlin man on his own is a potential source of trouble. The patrols ensured that men who had drunk too much were despatched, or dragged, back to barracks.

Unable to intervene or protest, Jack and his mates were aware that the French used to march the Arabs down a rocky gorge in the mornings and shoot them. Every morning! He was never told why this was happening, but was quietly informed that they were 'dealing' with thieves or informers. The British Army was unable to intercede as a new war with the French soldiers would not have helped. However, those actions by the French caused huge problems for many years after the War.

It seems that there was a lot of theft. At night the vehicles were parked up. The locals would creep in and either tie the soldiers' boot laces together as a prank or steal boots from them while they slept. They were clever enough to steal them without waking the wearer up. People stealing with skill seem to earn respect rather than disdain.

The troops would watch the progress of the war by reading the Union Jack newspaper that showed maps with advances marked on them. The feeling, by now, was one of reserved optimism.

There seemed to be a sense of relief as the main offensive in North Africa began to show positive results for the British and Americans. Lots of big bands, mostly American, and good vocalists performed to entertain the troops. In the NAAFI, an unknown black man came in and started to play boogie-woogie on a piano, the ultimate dance music of the day. Jack had never heard that style of music before and was captivated. The contrast between an East End of London pub pianist and a professional boogie-woogie player must have been immense.

The locals would sell eggs to the troops. Jack and the others would boil those eggs on heated anvils over a fire. Idiotically, one of the soldiers filled his can with petrol, put his eggs in it, and put the can on the makeshift stove. Of course, it blew up. He got sent home because he was so badly burnt.

Not to be too far from horses, Jack was loaned two Arab stallions in North Africa. He, and a soldier from Birmingham, rode them. Sadly, Jack's horse was not the best because when it ran it would hit its two back legs together. But, it was a horse and provided an escape from the War for a short period!

One day Jack was wandering up a street with Ernie Brooker in the market area. They heard the roaring buzz of an aircraft engine. They looked to see what it was and it started strafing with its machine gun. They leapt into a shop doorway, hid and waited until it was safe to carry on shopping.

Like the Grand Old Duke of York's Men, the troops had to march up a hill for meals and then march back down again, but this was in the intense heat of North Africa. One day, Jack and his mate sat on a bench to rest. An officer came along and put them on 'jankers'.

That is a formal charge that was entered into a charge sheet and they were confined to barracks for seven days. They were also given tedious tasks to do to enforce the humiliation.

So, on the second night they had to go to the sergeants' mess to make sandwiches. After their work on engines, they did not wash their hands. There were copious amounts of grease on the sandwiches, but nobody seemed to notice, and if they did they did not complain.

While they were confined in the kitchen, their other mates were passing bottles of beer through narrow gaps in the windows. Getting away with something was fun. Being tightly trained and with such an immense pressure for discipline, it appears that 'getting one over' on authority made it more worthwhile.

Moving closer to Tunisia, the REME troops, 750 base workshops, were moved to Constantine, half way between Algiers and Tunis They were transported in railway cattle trucks pulled by a train engine. His job was, again, reconditioning engines as a sheet metal worker.

The War was progressing elsewhere. In July 1943, Mussolini was deposed and Italian soldiers in Sicily were surrendering and in the August, British and American troops took Sicily. The campaign into the

Italian mainland had started.

Naples was the first Italian city to rise up and rebel against German occupation and the citizens had freed it by 1st October 1943.

The first Christmas away from home was difficult. Traditionally this was the time that would be spent with family and friends. Jack's son was a month old when his Christmas message was sent to his wife and baby, it was on a piece of paper. No glittering Christmas cards but a sincere message, nonetheless.

In January 1944, following the invasion of Italy by British and American troops, Jack was moved to Naples by ship. They slept on the ship overnight and drove the trucks to a town called Pozzuoli as part of the advance forces.

Pozzuoli was a much bombed place that started its life as a Greek colony before being taken over by the Romans and renamed in 194 BC. Laying almost equidistant between Rome and Naples, it is famous for being the place where another Sophia, this one Loren, grew up. Being an ancient town, it had many historical artefacts.

As if to express its indignation at the loss of life and the intense bombing in Italy, a few months after he arrived, Vesuvius erupted. Like a sleeping giant, it growled for a few weeks beforehand and then, on the

18th March 1944, the volcano that buried Pompeii exploded. It rumbled on and on. There were lava fountains and mixed explosions until the volcanic activity subsided on the 29th. While it was erupting the smoke could be seen from Pozzuoli and the red glow in the sky could be seen at night.

It has been quiet ever since. It resumed its sleep, and as if people want to call its bluff, many houses have been built in the area.

The work continued. There were always vehicles to be mended and restored; this time from the advancing American and British fronts.

Anniversaries were reminders of what was missing rather than what was there. At home his son was growing bigger without him and his still new wife was missing him.

Telegrams from home were short and to the point and censored with a razor blade to remove the place where the troops were.

Then, in April 1945, both Hitler and Mussolini died. Hitler at his own hand and Mussolini at the hands of Italian partisans.

Peace came to Europe in May 1945. Jack celebrated by going to the sergeant's mess, the only place that had beer.

After the War had finished some troops were moved on to Austria but Jack had to stay behind. He was asked if he wanted to go to Egypt, which he did not want to do, although his friend Ernie Brooker went. They had first met in Catterick and then again in Aldershot. They were both on the Windsor Castle when it sank and they stuck together until he went to Egypt.

In the office before he came out of the Army, he seemed to spend most of his time playing table tennis.

His dream of flying in a Lancaster bomber came true when he flew back from Italy for a month's leave after peace was declared. Full of curiosity, he went into the gun turrets and saw the world from a totally new perspective. That was the first time he had flown.

This was the only time he came home during his overseas service. He met his son for the first time when he was 2 years old.

He could not travel back by plane because the weather was bad so that they returned to Italy by train. That involved them having to stay in Dover Castle overnight. Better than the barracks he was used to, Jack was able to live like a king, for one night. As if to add to the sense of a post war holiday, they slowly travelled through France, then Switzerland back to Pozzuoli.

He stayed there, working in the office with two office girls, an officer and two other men. Italian women worked in the workshops. After working they were grubby and they used to strip off to wash. Nothing more has been added to that story!

The task was now to bring the workshop vehicles back to England. They used German drivers, ex-soldiers who wanted to return home. They travelled through the Brenner Pass, through Germany to Hamburg and then the trucks were shipped across to Hull. The German soldiers were left at Hamburg.

Jack was demobbed on the 30th October 1946, the month that ten high ranking Nazi war criminals were hanged at Nuremburg.

Jack and his best mate in the Army

The SS Windsor Castle after she was refitted for troop transport. She started life with four funnels.

The SS Windsor Castle shortly before she sank.

(A transcript of the ship's Master's Report of the attack and sinking is given in Appendix I)

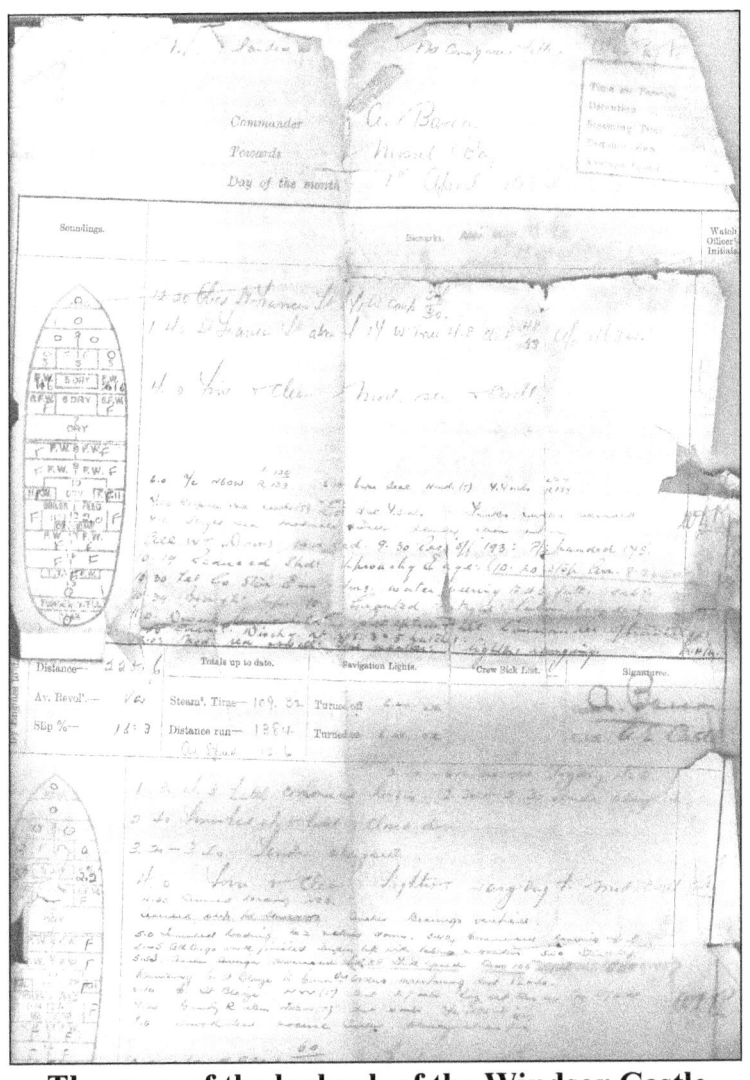

The page of the logbook of the Windsor Castle that Jack found on a beach in Algiers

Jack, on the left, with more mates.

Jack on the cattle truck that transported the troops to Constantine.

Sophie's War

Sophie was called up before her aunt because she wanted to do office work and she went into Pay Roll.

On the day she went into the RAF she got off a train, got onto, and off, a bus and walked in to the War as an active service woman now transformed from being a civilian.

She was given her service number, sent to the store room where she was measured for her uniform, her hat and shoes.

Later she was given her uniform in a kitbag and she changed into a WAAF. She was allocated a hut to live and sleep in

Sophie had not told her parents that she was enlisting. She assumed her gran told them in a letter! Her father was against her joining up. He would have been furious. For some unknown reason, he thought it was beneath the family to have a daughter in the Forces! Ironically, the day she joined the Air Force was the 29th November 1941, her father's birthday.

She started her training in Gloucestershire and then continued in Morecombe. At first she stayed in a house, but when bedbugs were found there, she moved into a hotel.

After worrying about the discipline imposed by her father, she found the regime of the Air Force more than acceptable. At least she was free from being ordered around <u>all</u> the time. She now had some leisure time when she could please herself.

She did not find a boyfriend; she was just happy to be away from being under her father's thumb. Strange to say that as her father had his right thumb shot off in the First World War. As mentioned before, he had joined up at the age of fifteen and perhaps after seeing how men reacted with women, he became ultra-protective of his daughters.

One First World War song was, "How Ya' Gonna Keep 'Em Down On The Farm, After They Seen Paree?" He had probably seen enough horror in, and beyond, the trenches to make him think about the bad things that can happen to innocents.

He loved his daughters and wanted nothing bad to happen to them. On his gravestone the word MIZPAH shouts loud. The usual meaning given is, "May the LORD watch between you and me when we are absent one from another."

However, a more relevant and direct meaning for the father of seven daughters is in the extended version, "Therefore its name was called Galeed, also Mizpah, because he said, 'May the LORD watch between you and me when we are absent one from another. If you afflict my daughters, or if you take other wives besides my daughters, although no man is with us--

see, God is witness between you and me!'"

After her initial training in Morecombe, she was moved to Penarth at the School of Accounts and carried on being taught bookkeeping.

She was transferred to Lewisham, even closer to the German bombs. She had to share a room with another girl who became a friend. They would go to the town and buy sweets or they would go into London to buy chestnuts.

Moving, step by step into more and more dangerous locations, she arrived at RAF Manston to work for Fighter Command. This controlled Spitfires and was subjected to constant and horrendous bombing. Due to its location at the East of the country it was involved in most types of operations from fighter planes to rescuing damaged returning bombers.

It was an airfield that did its utmost to shoot down German bombers before they got to London so it was a major target.

It was also where Barnes Wallis developed the 'bouncing bomb'.

She worked on Pay Rolls and if an airman went missing, the ledger was marked as "posted". The real meaning was never thought about. The idea of death in action had to be erased from consciousness.

Her gran wrote to her one day and told her that her father had not been well and she thought it would be a good idea for her to visit her parents.

She got leave and went home. There was a dance on in Thornford. Jack was there. They started going out and they arranged to meet on the Sunday morning. They went to Bath on a train to watch a film. Jack was home on leave and, at that point, they decided to write to each other, which they did. As their relationship progressed, Sophie would go up to London to meet him and they would see each other as often as they could.

In the April of 1942, they decided they would both go home on leave and she stayed with his parents. Jack asked Sophie to go to visit his brother, Fred, and his wife, Jennie, in Hastings. She was expecting her first baby. Jack was chopping wood and he said he would like to get married. Sophie asked him, "Who to?" and he said, "You, of course! But if you don't want to marry me, you needn't, but I am asking you." Sophie accepted. After they had been to Fred and Jennie's place, on the next leave they did a lot of running around making arrangements.

There is a nice story about Fred, the truth of which came out after the war. As already mentioned, he was a farmer. One day a Messerschmitt came hurtling out of the sky with its machine guns rattling. Fred was sitting on his tractor and a bullet passed through his trilby hat just missing his scalp. On the strength of that story he was bought drinks galore. He loved his whisky.

After the War it transpired that, in reality, a gust of wind had blown his hat off and into the grass mower he was pulling for haymaking. The pointed end of the cutting blade made the hole in his hat. He thought it would be insensitive to have corrected his original tale at the time! Like the whole family, Fred had a wonderfully wicked sense of humour.

Sophie used to catch the milk trains back to London after meeting up with Jack in Haverford West. If there was an air-raid on it would stop in a tunnel if possible.

Once, they went for a walk and an old boy said to be careful because the milk train would come along shortly. It was half-past three. They asked what time it came and he said, "About seven o'clock." Milk trains were those slow things that chugged along the line and stopped at every station to collect full milk churns for use in the cities and towns at their destination. The journeys would seem to last for ever.

There was a time when they were both on the same train at the same time without knowing it. They had not arranged anything because neither knew if they had leave. They were not married at that point. Sophie got out of the train and saw Jack walking down the platform towards her. They had been on the opposite ends of the train. They were both going home on leave.

After their leaves, they went back to the various places where they were stationed. They wrote to their parents. Sophie got a letter from Jack's sister, Dorothy, that said, "We have discussed it with your parents and you <u>will</u> be getting married on the 29th August in Thornford Church, and Betty and your sister Bobby will be your bridesmaids and everything will be arranged for you to be married."

Jack and Sophie got married because they were in love. He would be on leave before going away. It seemed a good time. If they got married they could save money because Jack would get a 'married allowance'. Jack was twenty.

Sophie's father was not keen on her marrying Jack, he tried to put him off and told Jack that he did not think they should be married because she had been out with a lot of other boys and had been 'playing around'. This was totally untrue.

He also told him that he had come home one day and found her 'canoodling' with a Scotsman on the settee. She had never been out with a Scotsman in her life. She only found out that this had been told to Jack after her father had died.

On the few occasions she was on leave, she and Betty

went to Bournemouth and Sophie bought silk underwear for the first time in her life.

Jack's mother had a friend who said that she could borrow her daughter's wedding dress rather than get married in uniform. The dress was the most beautiful shade of blue and it fitted as if it had been made for her.

They went into Yeovil and for bought, for the bridesmaids, smoky pink dresses with brown trimmings, headdresses and shoes.

Jack and Sophie were now were meeting as, and when, they could. The bombing in London was really bad but they managed to still see each other.

When they married they were lucky. The lady in the village shop gave Sophie all their flowers for her bouquet, the baker in Sherborne gave them their wedding cake and it had a little silver doily on the top because there was no sugar for icing. Jack's father's boss gave them some home-made wine and a man in the village had just enough petrol to take her and her future father-in-law to the church in his little car.

Why did she go to the Church with her father-in-law rather than with her father? Her father had always said that he would not give her away to be married and that he was not going to be at her wedding. So she asked her father-in-law to give her away.

They got to the Church and her father was there dressed up in his best suit. He went up to Jack's father and said, "O.K. Lou. I'll take over now, if you don't mind." He did mind! Sophie was blissfully happy and did not know that all this was going on.

One of the girls Jack went out with at Woolworths used to chase him and would invite him for tea. Later on, they shocked her one day. After getting married on a Saturday, on the Monday they went into the town and bumped onto her. "Hello Jack, nice to see you." She ignored Sophie. She asked Jack if he would like to go over for tea. Jack replied, "I'd love to if I can bring the wife!"

A week later, on the following Monday Jack was twenty one and he had to go back, off leave.

They both went up to London and they managed to see each other as often as they could. One night Jack travelled to London from Aldershot with his mate, Ernie Brooker. He did not have a pass. Sophie was waiting for him at Waterloo station. She was stopped by a sergeant. He asked her if she was waiting for somebody and she told him she was waiting for her husband. He asked where he was coming from and she told him, Aldershot. He asked what unit he was in and she told him the REME. "Oh," he said, "My name is RSM Massey." Sophie told him Jack's name. When Jack got off the train she told him who she had met and Jack and Ernie picked her up and carried her, at full speed, through the station. He was their RSM. They disappeared down the underground and hid.

They got a leave together just after the Christmas but in the March of the next year she sneaked away and managed to get to Aldershot to see Jack who was on embarkation leave. All the soldiers were assembled in the square in Hook, Hampshire. They were ready to go off. That was the last time she saw him for three years. But, on the last day before he went away, Sophie had managed to become pregnant with their eldest son.

Jack went and Sophie got severely hauled over the coals by her officer-in-charge and told that the powers-that-be would get Jack sent away for her misbehaviour. She burst into tears and told them they were too late because he had already been sent away.

She thinks that she felt a bit sorry for her and out her on seven days 'confined to barracks' and put her on light duties because she said she thought she was pregnant. She was supposed to go to the cookhouse and peel potatoes. The chaps there felt sorry for her and said that if she darned their socks, they would peel her potatoes. That is what she did. She darned socks.

Just after that, in the May, she wrote to Jack and told him she thought she was pregnant. He wrote back and said that the night that they were torpedoed, it flashed before him that he had left her with a baby. She did not read the word 'torpedoed' or any other information because the censors had covered the words with a blue pencil. The people who read the letters either did it with a blue pencil or cut out the piece with a razor blade. Anyway, she knew something had happened because in her letter she had asked him if there anything he wanted. He replied to say he would be grateful for razor blades and various things because at that moment he was walking about in his socks. He had no boots.

The alarm bells in her head started ringing. She wrote to him and said that she would ask him questions and when he replied he would say "In answer to your questions 1, 2, and so on, the answers should be 'yes' or 'no'." She asked all manner of questions. Her letters to him were not censored, but his to her were. So in the end she found out he had been torpedoed. She knew he had sailed so she could ask him if the boat had been hit. He wrote back, 'yes'. She, at least knew that despite a near miss, he was safe.

Sophie and other WAAFs (Women's Auxiliary Air Force) who were stationed at Manston were billeted at the nearby Ursuline Convent in Westgate on Sea for their greater safety from the bombs.

The raids went on night after night. The sirens kept going off. It was not a nice time for her. So much time was wasted in the air raid shelters. There was only once when she did not go to the shelters when a raid was imminent.

One night, she was with her friend, Dorothy. The siren went and Dorothy said that she felt they should not go down into the shelter that night, she did not feel comfortable. They crept out into the grounds and hid under a hedge. An air-raid warden came along and asked them what they were doing. He told them to go to the shelter because he was not happy that they were in the open. They were insistent that they should stay where they were, nonetheless.

Later on, a bomb landed about twenty feet from where they were and the earth and dirt showered down on them. After the raid had ended they saw that there was a huge crater. The shelter where they should have been in had taken a direct hit and all the people in it had all been killed.

In that shelter there had been the mother of a young woman who was stationed at Bedford. Sophie was asked if she could exchange postings on compassionate grounds so that the young woman could go to Manston to be closer to her family.

Sophie thought she had nothing to lose by going to Bedford. She was pregnant anyway so her time in the WAAFs was limited. So she went to Chicksands Priory. That was in the April.

She went home to have her baby in the autumn. There was no room at her parent's house because she had six sisters by then so she went to stay with Jack's parents at Lake. Being on the farm was probably felt better for her than being at her parents' house, anyway.

Before her baby was born, Jack's father had an argument with the farm manager who took over from Mr Perham. Jack's father was a gentle soul who rarely, if ever, lost his temper but on this occasion he must have been threatened and used his best metaphorical put-down…

"I have lived too near the woods, too long, to be frightened by a bloody owl like you!"

And with that, he resigned.

The family moved to a farm in Wiltshire for a while but the practicalities of having to walk miles from the house they were supplied with to the fields made it a nightmare. Whilst at this farm, Sophie gave birth to her son, Vivian, who later changed his name to Victor. The delivery was in the house with a midwife and Sophie's mother-in-law in attendance. It was painful and slow, but she was now the mother of her own child.

Jack's father applied for another job that was advertised in the newspaper, this time near Weymouth in Dorset.

This farm was a large one. The fields stretched from over the top of the White Horse carved into the side of a hill, to the cliffs at Osmington. The White Horse carrying its rider George III, had a mixture of stories surrounding it. Some said it was carved to commemorate Royal visits but that the 'Royals' were offended by it showing the King riding away from Weymouth, thus suggesting that he was not welcome.

Other stories told that it was built to occupy the engineers' time whilst waiting for Napoleon to invade. Come what may, this horse and rider was, and is, a huge landmark. The hill to the fields above the horse is steep and horses were still used to drag wagons up for hay, wheat and other crops. Oddly enough, the white chalk was not covered or painted during the War. It seemed to be an obvious navigation point for the Luftwaffe pilots, but the mass of Portland Bill must have been more significant.

The cliffs a mile or so away to the coast were fenced off to civilians. Gun placements abounded. They pointed out to sea to attack incoming aircraft and could be used to shoot at any enemy warships that might approach.

In the build up to D-Day (6th June 1944), chaos reigned. Weymouth Bay, plainly visible from Osmington, was packed with Aircraft Carriers, Battleships, Destroyers and Motor Torpedo Boats, waiting for the word.

There was so much activity around that Sophie and everybody suspected something was about to happen. The roads were completely blocked by American tanks and trucks. Smoke from their exhausts choked the air. During the build up to D-Day she could not get over the main road so could not get out of the village.

By then Sophie had her little boy. He was now six months old. The protective maternal drive for his mother was to keep him safe in a world that was hostile.

She could see the temporary harbour being built in the bay and she heard the bombers going over and then back again. There was a perpetual roar of engines. She would count the bombers out and back in again. Unlike the planes counted by the Falklands reporter however, the number on the way back was always less than that going out. This was part of the effort to reduce the opposition to the beach landings that would come.

This manic activity went on until the end of the War.

While her mother-in-law looked after her son, Sophie had to help to pick potatoes and then she had to collect the small ones up to use for meals and to feed the pigs. It was the same during harvesting. After they had cut the corn, she had to glean the fallen ears of wheat for chicken feed. She was an added hand to the labour that was necessary to maximise the crops. This left Sophie feeling that Jack's mother was having more influence on her son than she did, but she was now in her mid sixties and made a better child-minder than labourer.

The farm employed Jack's father, Harry and Bill. The other brothers had moved away when they lived in Thornford.

On the farm, Jack's father always kept six cockerels which would be fed up and sold at Christmas to raise money for presents. The family would all have to sit around a zinc bath and pluck them. They would tie scarves around their faces to avoid inhaling the dust from the feathers.

Sophie used to go into Dorchester to draw her Army pension. Just down the road from the Post Office was a butcher's shop where people could buy Bath Chaps, the pig's jaw, off ration. They would buy fish from MacFisheries.

As had happened in her parent's house, washing was done in a big vat called a copper and water squeezed out with the mangle. The work was hard. Sophie thought that she had, in her own words, "Swapped, Old Nick for the Devil".

Jack's mother was the local unqualified midwife and helped out with births as, and when, required. But she was very strict, maybe domineering to Sophie. She would upset Sophie and then, the next day, would buy sweets to say sorry.

The War progress was listened to on the radio where the family would gather to listen as families today watch the television.

The good news came that the War had ended but the celebration was quiet and restrained. No parties, just a sense of relief that Jack might be coming home soon.

Some of her sisters, by now, had husbands who were also going to be relieved of the burden of War.

Minnie's husband was involved in maintaining electricity supplies, a 'reserved' occupation. Like the farmers, reserved occupations were those that were for the essential duties in Britain; food, power, fire-fighting and so on.

However, some had to wait for the fall of Japan. Florie's husband, Frank was a Japanese prisoner of War who was badly injured and tortured. Bobbie's (Roberta) husband was in the Pacific with the Navy.

Jack came back to England for his 28 day leave after the War ended, but had to find his parents, wife and son. In his time away they had moved to the farm near Weymouth. No more the comfort of Thornford and his friends.

His trip was delayed after the Lancaster landed in Peterborough because one of his mates was ill. They stayed with him for a day or so before he could set out to find his wife and son.

He got to the village and asked a man where the farmhouse was. He walked down the road with him and showed him. He walked back to the pub and as he went he shouted, "You didn't even buy me a bloody pint for showing you where it was." Later on, Jack saw him in the pub and said, "Here you are. I'll buy you a bloody pint, now. I've just come back from the War and you useless bastards are just here, doing nothing."

The family dynamics were strange. Sophie's son thought his father was a photograph. Every night Sophie would ask him to kiss a photograph of Jack, 'goodnight', so when Jack returned he did not consider him his father because the photo was! So the poor lad thought a photograph was his father and that his grandmother was his mother because his she, by that time, had taken on the role of surrogate mother so that Sophie could work on the farm.

The visit was good. Jack had met his son, at last. Sophie was ecstatic to see her husband, but the stay was too short before he had to return. Yet this time it was not for War but for the settling in of Peace. Still, Sophie had to wait for over a year before Jack came home for good.

Sophie and her fellow WAAFs

Jack and Sophie's wedding

Jack's father, Louis, with his chickens.

Jack's mother with Viv

Sophie with Viv

Home Thoughts From Abroad

The Telegrams

Best Wishes For
Christmas and
the New Year

To My Darling Wife and Babs

    I only know that suffering,
    And pleasures very few
    Have made me realise that I
    Have lots of love for you

From Your Ever-loving Husband Jack,

Lots of love darling

B.N.A.F

25th DECEMBER 1943

NEWTON 25 JANUARY 1944

VERY HAPPY TO HEAR FROM YOU DEAREST
THOUGHTS ARE WITH YOU

FONDEST LOVE

SOPHIE

BRADFORD ABBAS 29 FEBRUARY 1944

VERY HAPPY TO HEAR FROM YOU DEAREST AM FIT AND WELL I WISH WE WERE TOGETHER ON THIS SPECIAL OCCASION MY BEST WISHES FOR A SPEEDY REUNION

FONDEST LOVE AND KISSES

SOPHIE

     (received) 14 FEBRUARY 1944
TISBURY 8FEB

LETTERS ARRIVING REGULARLY VERY HAPPY TO HEAR FROM YOU DEAREST AM FIT AND WELL FONDEST WISHES FROM ALL OF US.

SOPHIE

SOPHIE 22 AUG 1944

LOVING BIRTHDAY GREETINGS TO DADDY MY THOUGHTS AND PRAYERS ARE EVER WITH YOU.

SOPHIE

CHRISTMAS 1944

To my Darling Wife, and Vivian

Your Ever-loving Husband, Jack

(This telegram has a sketch drawing of the Coliseum in Rome on it)

Christmas 1945

To My Darling Wife
Wishing you all the Very Best Wishes for Xmas darling.

From your Ever-loving Husband, Jack

The Postcards

On his way home at last, Jack sent postcards to Sophie to keep her informed of his progress. The following are transcripts of the messages.

Postcard of Milan, 19 June 1946
Darling Soph

At Milan, arrived tonight, no idea when leaving yet, hope it's tomorrow. Will write letter if we stay here long. Journey alright so far. Just longing to reach England, darling. Lots of love Soph, will soon be with you now.

Your ever loving husband, Jack. XXX
To Viv with love. XXX

Postcard of Rimini, 18 June 1946
Darling Soph

At Rimini going well, 200 to Milan, arrive tomorrow, Wednesday. Tons of love.
Your loving husband. Soon be with you darling.
Jack XXX

Postcard of Lake Como, 25 June 1946
My dear, Darling Wife.

Well darling I'm afraid we are still stuck at Milan. We were supposed to leave today but now it's jumped back to definitely Friday. It's getting a dead loss, and knocked our chances of being earlier, a bit silly. Well, darling, I will send another card Friday to let you know if we are on our way or not.

Your loving husband, Jack. XXX

Postcard of Milan, 27 June 1946
My dear Darling Wife,

Hello darling, definitely leaving tomorrow, Friday, for Hamburg. Will send postcards to let you know progress. Hope we are lucky at Hamburg for ship, darling. Hope you and Viv are keeping fine darling, will soon be with you. Know our great day at last, all my everlasting love, darling.

Your husband Jack. XXX

Postcard of Frankfurt, July 1946
My Dear Darling Wife.

Sorry not to have been able to write before darling, no place to post it on journey. But we have arrived at Hamburg tonight, don't know when leaving but be prepared for the telephone call, soon be with you now darling, our great day.

Lots of love darling.

Your loving husband, Jack. XXX

Postcards

More postcards

Winston Churchill

May 8, 1945 London

"My dear friends, this is your hour. This is not victory of a party or of any class. It's a victory of the great British nation as a whole. We were the first, in this ancient island, to draw the sword against tyranny. After a while we were left all alone against the most tremendous military power that has been seen. We were all alone for a whole year.

There we stood, alone. Did anyone want to give in? Were we down-hearted? The lights went out and the bombs came down. But every man, woman and child in the country had no thought of quitting the struggle. London can take it. So we came back after long months from the jaws of death, out of the mouth of hell, while all the world wondered. When shall the reputation and faith of this generation of English men and women fail? I say that in the long years to come not only will the people of this island, but of the world, wherever the bird of freedom chirps in human hearts, look back to what we've done and they will say "do not despair, do not yield to violence and tyranny, march straightforward and die if need-be unconquered." Now we have emerged from one deadly struggle-a terrible foe has been cast on the ground and awaits our judgment and our mercy.

But there is another foe who occupies large portions of the British Empire, a foe stained with cruelty and greed, the Japanese. I rejoice we can all take a night off today and another day tomorrow. Tomorrow our great Russian allies will also be celebrating victory and after that we must begin the task of rebuilding our hearth and homes, doing our utmost to make this country a land in which all have a chance, in which all have a duty, and we must turn ourselves to fulfil our duty to our own countrymen, and to our gallant allies of the United States who were so foully and treacherously attacked by Japan. We will go hand and hand with them. Even if it is a hard struggle we will not be the ones who will fail."

Jack and Sophie, Post-War

War had transformed the world. There had been a metamorphosis from the peaceful tranquillity of the pre-War days into the start of a contradictory drive for acquisition in a period of austerity. People had seen a greater picture. They had seen 'foreign' countries and different cultures, albeit with weapons in hands and rubble underfoot.

Soldiers, sailors and air force men and women had fought, not only for the defence of Britain, but for the liberation of the rest of Europe and the Far East.

But the new war, the Cold War, against Communist countries reflected a paranoid feeling. The West wondered if Russia, the wartime ally, had the knowledge to make atomic bombs.

India was being divided into two parts based on religion, and Palestine would have its bloody birth.

The war had spawned change.

Although there was a hope, perhaps an expectation, that rationing would come to an end shortly after the end of the War, it remained in place until 1954.

Canning was a more practical way of providing preserved foodstuffs and the regular use of tinned fruit and meat was born and would partially replace the need, and opportunity, to buy and consume fresh food. The Country needed convenience and speed.

Life on the Osmington farm became busier as pressure was applied to increase production even more. Demand during the War was high enough, but with returning soldiers it had become massive. Chemical fertilisers supplemented the more natural use of what we would call today, organic, methods.

Relationships had changed from that of having somebody who loved you despite the worry that you might never came back, to the reality of sharing every moment in the new ambitions for housing, food and work. If marriages and partnerships were to last with happiness, they had to be sound.

Jack arrived home with his demob suit, a new shirt, pants, socks and shoes plus a trilby hat which he never wore.

Jack and Sophie were now living together for the first time. They had been together for only a short time in total during the past 4 years. Rather than a stranger returning after all that time, Jack was welcomed by his old and new families. He was given three months paid leave in which to re-establish a life both in England and in 'civvy street'.

Happily, his relationship with Sophie picked up from where they had left it. It was bliss for both of them to be together and wonderful that there was now a son in the new scheme of things.

Jack was back in Dorset again, this time on the South coast rather than in Lake. This time the farm had just his father, Harry and Bill to work it. His mother, wife and son were also there but it was not a place that Jack could learn to adapt to as he had developed chronic hay fever. Although he was free of it before he went overseas, on his return it became debilitating. It prohibited him from getting involved during haymaking or working during the summer with the wheat, oats and barley.

The people were noticeably older. Jack's father was now 66 and found the work harder. His broken legs slowed him down on the steep hills up to the fields. Yet, this was the Post War period when pressure was being applied to increase crops for a Nation that was being rationed.

Jack worked as a handyman and driver for a while. A major task was to learn how to drive on the left of the road. He learned to drive in Italy, in much the same way as he had learned to ride. The soldier driving a big truck slid across the seat, took his hands off the wheel and told Jack to carry on. This meant that he became familiar with driving on the right. Luckily, traffic was lighter then than it is today.

Sophie, ever ambitious for her family, saw an advertisement for a panel beater. Jack was used to working with broken metal to the point where he could repair it with barely a mark showing.

Jack got the job that enabled them to move off the farm. He worked locally for a while but his need to earn more for his family was growing.

They saw an advertisement for a job in Bristol. A man called Mr Lewis, from Windmill and Lewis, travelled down to interview Jack.

He sat down and took his Trilby hat off. He stopped before passing it to Jack and said, "So, you are telling me you could take a car as crumpled as this…" He proceeded to squeeze the material into a tiny ball. "…and smooth it out." He gave it to Jack.

Jack restored the hat to its former shape it and passed it back. "Of course I can." After talking terms that included accommodation he was offered, and accepted, the job.

The three of them moved to Bristol, a place of great camaraderie and fun, even in those years of austerity. Because everybody was in the same boat, nobody worried about too much. That generation had survived a war.

The evidence of the bombing was everywhere. Buildings flattened to rubble abounded in Bristol.

Sophie found she was pregnant again at the beginning of 1947 and insisted on having the baby in a hospital rather than a home delivery as she did with her first son.

Jack was now in the motor body repair business and when he was offered a job in Yeovil a few years later, the family, now the four of them, moved to the next stage of life.

They moved back to Bradford Abbas where Sophie had come from and Sophie's mother looked after her youngest son at lunch breaks from school while Sophie worked. The wheel had turned.

Jack cycled the three miles to work each morning and the three miles back, six days a week until the journey became too tiresome for him and they rented a house in Yeovil.

As the intention has been to relate a story about two people before and during the War, the story can rest there. It is enough to say that they returned, yet again, to Bradford Abbas to rest from a long life of hard work, and they still live there today.

Jack outside Osmington farm

APPENDIX I

The following is the account by the Master of the S.S. Windsor Castle, the ship that Jack was on when it was torpedoed in the Mediterranean on her way to Algiers. It was built in 1922. Initially she was designed to carry 234 First Class, 362 Second Class and 274 Third Class passengers with a crew of 440, a total of 1310. On the day she was torpedoed she was carrying 3055 people.

Note: Explanations of equipment details have been added by the author in square brackets and in bold

Reproduced with the kind permission of the Naval Historical Branch of the Ministry of Defence.

CONFIDENTIAL   TD/139/1806
<u>7th May, 1943</u>

<u>SHIPPING CASUALTIES SECTION, TRADE DIVISION,</u>
<u>REPORT OF AN INTERVIEW WITH THE MASTER, CAPTAIN J.C. BROWN,</u>
<u>S.S. "WINDSOR CASTLE"</u>   -   <u>19.141 g.t.</u>

CONVOY ; K.M.P. 11

<u>Sunk by 1 torpedo from Aircraft on 23rd March, 1943</u>

<u>All Time are B.S.T. - 1 hour for G.M.T.</u>

CAPTAIN BROWN :

We were bound from the Clyde to Algiers carrying troops and equipment. The ship was armed with a 3" H.A. **[high-angle anti-aircraft gun]**, two 12 pdrs. **[anti-aircraft guns]**, 1 Bofors **[a heavy anti-aircraft gun]**, 10 Oerlikons **[20 mm automatic weapon]**, 4 F.A.M's **[possibly flares]**, 4 Pig Troughs **[radar]**, 4 P.A.C. rockets **[parachute and cable anti-aircraft rockets]**, 3 depth charges; paravanes **[devices that were deployed from the bow to disable mines]** were fitted, and we carried a balloon but this was not flying at the time. The crew, including 16 naval and 13 army gunners, and a Lieut. R.N.V.R. (as Gunnery Officer) numbered 297, and we carried 2,758 troops – as far as I know, although I learned later that a further 3 troops are missing. All confidential books, including wireless, were ready to be thrown overboard in weighted boxes, but were left to sink with the ship. We carried about 12 confidential bags of mail which were stowed in the Specie Room; this room was not damaged by the explosion, and all bags went down with the ship; there was no possible chance of compromise. Degaussing was on.

2. We sailed from the Clyde on the 16th March, joining up with convoy K.M.F. 11, later taking up the position No. 31. We proceeded until about 1900 on the 19th March when we sighted a Focke-Wulf Condor a long way off to the Eastward. After a short time she disappeared, but we knew she would report the

convoy. We carried on, steering evasive courses, until 1400 on the 20th when the Escort hoisted a signal "Formation of enemy aircraft approaching" and giving the bearing. The whole convoy went to action stations, but nothing developed and a signal was then made "Friendly aircraft approaching". A little later another signal was received "Hostile aircraft coming in", which was again followed by "Friendly aircraft in sight"; we carried on but once again a signal was made "Hostile aircraft approaching". Action stations were resumed, and finally a last signal stated "Friendly aircraft approaching". And the whole thing blew over, having lasted about 2 hours. The day was heavily overcast and we did not see any planes at all, so either the Escorts were mistaken, and were correcting their mistakes, or else both enemy and friendly planes were about, the first attacking, and the others driving them off.

3. We proceeded without further incident, zig-zagging all the time, until 0235 on the 23rd March, when, in position 37 degrees 27'N., 0 degrees 54'E., steaming at 13½ knots on a main course of 078 degrees (approx.) the sound of an aircraft was heard on our starboard bow. It was not visible, but suddenly my 4th Officer picked it up through the binoculars about 2 miles off, and watched it for a moment or two as it flew along the starboard wing of the convoy, passed over the destroyer which was on our beam, then flew ahead and was lost to view. We could still hear it flying round, apparently circling ahead of the convoy,

but no signal was made and nobody opened fire. After about 5 minutes, as nothing happened, I presumed it was a friendly plane and went to lie down, as we were nearing Algiers and I thought it was the air Escort for the convoy. The next thing that happened was a splash just ahead of the S.S "NEA HELLAS", leading ship of the second column. The "NEA HELLAS" was then about two points abaft our port beam in the next column, some 5 cables away. Immediately after the splash the plane came into view flying at as height of about 150 feet, it banked steeply away, and 10 seconds later, at 0240, we were struck by a torpedo. It was moderately cloudy at the time, but fine with good visibility and a full moon, there was a slight sea and heavy swell, with light airs.

4. The torpedo struck on the port side in No. 4 hold abaft the engine room. There was a terrific explosion, no flash was seen, but a column of water was thrown up as high as the bridge. All the hatches and beams of No. 4 hold were blown away, No. 4 hold flooded immediately, and as both bulkheads were shattered the water poured into the engine room and No.5 hold, all three compartments being flooded within 2 or 3 minutes. No. 16 boat was damaged by the explosion, and the ship took a list of 3 degrees to port, which remained until she finally sank. As soon as we were struck the "emergency" bells were rung, two white rockets were fired, the red lights switched on, and the sound signal made indicating that we had been torpedoed on the port side. A wireless signal was also

made, and the engine room telegraph rung to "STOP", but no reply was received as the engine room was flooded. The ship slowed down and the engines stopped automatically. Five or ten minutes later the S.S. "CUBA", the next ship astern of us, passed along our port side, and when off the port bow opened fire with all her guns towards the port side as the plane had circled ahead and was coming in for a second attack, probably knowing she had scored a hit, she wanted to finish us off with a second torpedo. Fortunately, the "CUBA" saw the plane in time to open fire, throwing up her "flak" at an angle of approximately 30 degrees; no other ship opened fire and the attack ended.

5. I sent the Chief Officer and Carpenter to ascertain the external damage to the ship, and the Chief Engineer to investigate the internal damage with the Carpenter's Mate. I then sounded the emergency alarm bells for all troops to muster at their emergency stations, which was carried out quickly and in perfect order. I then waited for the damage reports from everybody. Two air pipes leading into the after stokehold were fractured and were allowing water to enter from the engine room, so I sent Mr. Dickenson, 1st Officer, Mr. Bonham, 4th Engineer, and McNeil, the Carpenter's mate, to plug them. They worked efficiently and quickly, thus preventing all water from coming through; these three men were deep down in the ship, never knowing when the bulkhead might give way or the ship go down. The Chief Engineer

had closed all valves, checked up on watertight doors, and although the fires in the boiler room had automatically gone out, to ensure against risk of fire, the further precaution was taken of shutting off the fuel supply to the burners.

6. On learning that the ship was helpless, I considered that the troops should be got away as quickly as possible, as I thought we might get another torpedo, submarines having been reported in the vicinity by one of the destroyers. I therefore gave the order to abandon ship, and within 10 minutes every lifeboat, with the exception of the damaged boat, was fully loaded and clear of the ship; there were 19 boats altogether, which took off 1,100 troops, all being in charge of the Chief Officer. Nothing went wrong, we had practised boat drill many times during the voyage, and the whole operation was remarkably successful. All except 4 of the lower boats, and the motor boat, were ready swung out, the motorboat being too heavy to be kept on the falls all the time, and the other four being the lower of two pairs under davits. All davits were equipped with the ordinary rope falls.

7. Meanwhile, the troops assigned to rafts stood quietly at their stations until the lifeboats had left, as arranged previously, these raft parties then prepared to disembark. I decided to re-examine the ship to see if there was a chance of saving her, so I posted Officers in various parts of the ship to watch and report on the

situation, and learned that the water in the flooded compartments remained stationary.

H.M.S "LEAMINTON" and "Whaddon" were standing by, one of them signalled that she would take off the troops assigned to the rafts, and came alongside to do so. I therefore ordered the troops to cease dis-embarking on the rafts until she arrived, but several rafts had already got away. Most of these returned to the ship, and first "LEAMINGTON" and then "WHADDON" came alongside, taking off all the remaining troops with most of the crew, whilst I retained 35 Officers and men of the ship's company. I had requested the Destroyer to signal to Algiers, about 110 miles away, for tugs, as I considered there was a chance of saving the ship, and she reported that the signal was acknowledged at 0300.

I considered the men remaining on board adequate for towing purposes, or for such salvage work as might be possible, but later this number was increased to 100 by volunteers who deliberately returned to the ship to give assistance; in fact I had to ask the destroyers not to allow any more men to return to the vessel. All D.E.M.S ratings in charge of the Gunnery Officer and Sgt. Whittaker returned, all guns were kept fully manned.

8. I asked for reports of the situation to be made, by the various Officers and men I had posted, every quarter of an hour, and at first the position remained

the same, but two hours later, at about 0500, I was informed that water was coming into the tunnel aft and was beginning to increase in the various flooded compartments, but No. 6 hold was still dry. They had the emergency pump working on the tunnel compartment as this was not open to the sea and I wanted to raise the stern to bring the ship on an even keel if possible; she had now sunk 16 feet by the stern. An emergency diesel engine operates the emergency pumps, the after emergency pump was tried, but we found this had been put out of action by the explosion as it was situated in No. 4 hold. The emergency pump in the foremost boiler room was then started and this at first lowered the water in the after tunnel compartment, but the water in the engine room, together with Nos. 4 and 5 holds, started to rise slowly.

9. Meanwhile I continued to receive reports from the various parts of the ship every quarter of an hour, until 0945 when something I cannot explain happened, and a sudden deterioration of the position all over the ship set in. The water in the flooded compartments rose rapidly, and at the same time the water in the tunnel compartment overtook the pump and rose to the level of the watertight door in the lower deck. No. 6 hold, which had previously been perfectly dry, now had 6 feet of water in it which was rising quickly, and the water was coming into the stokehold from the engine room bulkhead at the level of the lower deck. The stokehold had been dry, as had the after boiler until

this time, but now there was 8 feet of water at least, rising so rapidly that we could not ascertain where it was coming in, but I think the bulkhead must have been fractured. The water was now only 3 feet from the 'B' deck in No. 5 hold, 4 feet from this deck in No. hold, and 6 feet from the deck in the engine room. I reckoned that the ship was filling at the rate of 800 tons an hour, and the Chief Engineer estimated it at 1000 tons an hour.

10. Three alerts were sounded against unidentified aircraft during this time, but all proved to be friendly after being challenged by the Escort.

11. There was no sign of the tugs, I understand they did not leave Algiers until 0700 – 4 hours after our signal for assistance; I realised the tugs could not possibly arrive before 1800. The question of towage by the "WHADDON" and "LEAMINGTON" seemed to me to be impracticable, as both were filled with troops and were required for anti-submarine patrol. We were 110 miles from Algiers, the nearest port, and 58 miles from Cape Tenez the nearest land. At about 1015 two more Destroyers came along to relieve, and "WHADDON" and "LEAMINGTON" left for Algiers with the troops, but the Captain of the "WHADDON" asked one of the relieving destroyers to accompany him as an Escort, as the loaded destroyers had little manoevrebility in the event of an attack. H.M.S "LOYAL" remained standing by. Just before she arrived, however, water was reported rushing along

the orlop deck alleyway which, until then had been dry, and pouring down into the after boiler room. The water level in the engine room had been above this alleyway from the first. There now seemed no hope of saving the vessel unless the salvage vessel arrived very soon, and there was nothing more we could do until she arrived. I considered it imprudent to keep my Officers and men on board any longer, so at about 1020 I gave the order for final abandonment of the ship. We were taken on board "LOYAL" to await the arrival of the tug and salvage vessel. The time was about 1030. I discussed the question of towage with the Captain of H.M.S "LOYAL" and gave it as my opinion that nothing constructive could be accomplished until assistance arrived and I could consult with the salvage experts. Moreover, the ship was to be steered by power as the emergency electric steering engine was now submerged. I expected the ship to sink within the hour, actually she remained afloat for a further 7 hours, but the tugs did not arrive before she sank.

12. About 1500 another destroyer arrived and, together with "LOYAL", they carried out a submarine patrol round the ship. This continued until just before 1700 when more destroyers arrived, including H.M.S. "ESKIMO", and at that time I was informed by the Captain of the "LOYAL" that a signal had been received from the Commander-in-Chief ordering the destroyers to take the vessel in tow. He asked if I would go on board "LAMERTON" to discuss

operations; I replied that I certainly would, but I knew the case was hopeless. I took my Chief Officer with me, but on the way over to "LAMERTON", as we were drawing alongside, her Captain shouted for us to return to "LOYAL". As we returned, I noticed another destroyer, H.M.S. "FARNDALE", alongside "WINDSOR CASTLE" putting men on board. I at once ordered the coxswain to put me on board also. We went straight to the ship, when I saw immediately that her end was near, as the water was rising rapidly aft and 'B' deck was now well covered, so there was nothing to prevent the after end from filling up completely. I therefore went to H.M.S. "FARNDALE", hailed her, told the Captain that the vessel was sinking rapidly, and to get his men off at once. This he did only just in time; the last man was sliding down a rope from the bow when the "WINDSOR CASTLE" went up on end and sank stern first, the time being 1720.

13. I returned to "LOYAL" which proceeded at once to Oran. I am proud to report that the behaviour of all my Officers and men was beyond praise. When I retained the 35 Officers and men on board, I had to order the remainder to leave the ship, and later, when a number would return, I had to signal the destroyers requesting the Captains not to allow any more to come back. I had the utmost loyalty and co-operation from everyone, the men worked willingly and eagerly in an endeavour to save the ship.

14. Had it been possible for "WHADDON" and "LEAMINGTON" to tow us after the casualty occurred, we should only have been able to proceed at about 2 knots, and could never have reached the land. I do not think it would have been possible without tugs, which never arrived. We had prepared a 6½" wire on the fo'castle head, which was kept on the reel, but luckily "FARNDALE" had not made it fast when the ship sank. It would have been impossible to slip it quickly and the destroyer would inevitably have been pulled down with the "WINDSOR CASTLE". We should in any case have required another destroyer to steer us, as the emergency steering engine was submerged. We could not use our cables, as we had no power on the capstan. We made the 6½" wire ready forward, and a 5½" wire ready aft (both wires being only 120 fathoms long) and we were as ready for towing as we could possibly be without power. The tugs usually have an 18" manila rope of 230 fathoms; even they could only have towed us at 3½ knots, so I doubt if the vessel could ever have been beached, had they arrived after the torpedoing.

15. We had 6 scrambling nets over the side for abandoning ship, which proved excellent; I hope to have at least 8 nets in my next ship.

POSTSCRIPT

A quick update is to say that they moved around, went back to Bristol, moved to Wantage in Oxfordshire. As if drawn to the start point, they retired in, and remain in, Bradford Abbas.

Ernie Brooker and Stan Foot kept in touch and were visitors. Jack is not too sure what has happened to Ernie, but he knows that Stan has died.

Jack, at 87, still prefers Thornford and Sophie, at 86, is happy where she is.

If you have not guessed it already, the heroes of this story, I am proud to say, are my parents. They have travelled a journey that is rich and full of happy memories and love.

A closing ghost story:

As mentioned, Sophie's grandfather looked after the graveyard in Sevenoaks. Whenever there was a funeral, they would always play the same record on the gramophone. He insisted that the music was never to be played at his funeral. When his turn came, his son forgot the instruction and placed the needle on the record. With an almighty whoosh, the arm was jogged across the record, scratching it so it could never be played again. That was Sophie's mother's father.

Other publications by the author include:

BOOKS

The Secret Language of Hypnotherapy
ISBN: **978-0-9550736-2-5**

Mind Changing Short Stories and Metaphors
ISBN: **978-0-9550736-4-9**

Short Stories and Metaphors
ISBN: **978-0-9550736-3-2**

AUDIO

Insight to Anger
Metaphorical stories about the destructive nature of anger and how anger can be overcome.

Stop Smoking
How to stop smoking easily and safely using hypnosis and breathing techniques.

Weight Control
How to reshape your body using hypnosis and visualisation. This method has helped thousands of people.

Fantasies and Dreams
Relaxation and positive thinking.

Self-Hypnosis
How to relax, visualise and set positive suggestions for yourself.

Animal Nature
Metaphors that relax and change the way you think based on animal stories.

Human Nature
Metaphors that relax and change the way you think based on stories about people.

Nature's Nature
Metaphors that relax and change the way you think based on stories from nature.

More information from: http://www.emp3books.co

www.ingramcontent.com/pod-product-compliance
Lightning Source LLC
Chambersburg PA
CBHW031358040426
42444CB00005B/340